Save Our Planet
An anti-nuclear guide for teenagers

Jim Eldridge
Illustrations by David Ayres

A Magnet Book

First published in 1987 as a Magnet original
by Methuen Children's Books Ltd
11 New Fetter Lane, London EC4P 4EE
Text copyright © 1987 Jim Eldridge
Illustrations copyright © 1987 David Ayres
Printed in Great Britain

This paperback is sold subject to the condition
that it shall not, by way of trade or otherwise,
be lent, re-sold, hired out or otherwise circulated
without the publisher's prior consent in any form
of binding or cover other than that in which it
is published and without a similar condition
including this condition being imposed on
the subsequent purchaser.

British Library Cataloguing in Publication Data

Eldridge, Jim
  Save our planet: the teenagers anti-
  nuclear book. – (A Magnet book)
  1. Nuclear energy – Environmental aspects
  I. Title
  333.79'24   TK9153

  ISBN 0–416–02512–9

SAVE OUR PLANET

An anti-nuclear guide for teenagers

Since Chernobyl the nuclear industry
has been trying to convince us
that nuclear power is safe.
This book sets out to explode that myth,
present the anti-nuclear case
and make positive proposals for
alternative renewable sources of energy.

This book is dedicated to Stewart Boyle,
and everyone else at Friends of the Earth,
who gave me so much information,
and who are still working so hard
to save our planet for the future.

Contents

	Introduction	page 9
Part One:	**The Nuclear Industry**	
	1: Nuclear Energy and Radiation	15
	2: Nuclear Power	22
	3: Chernobyl	32
	4: Nuclear Power – The Weapons Connection	44
	5: Nuclear Weapons	50
	6: Nuclear Waste	72
	7: Uranium Mining	81
Part Two:	**The Alternatives**	
	8: Energy from Renewable Sources	87
	9: Defence – The Non-Nuclear Alternative	101
	Glossary	104
	List of Useful Organisations	108
	Research and Further Reading	109
	Acknowledgements	110

Introduction

On 26 April 1986, at Chernobyl in Russia, a nuclear reactor exploded. Officially, 35 people have died at the time of writing, and doctors and scientists estimate that many hundreds more will die of cancers as a result of the blast.

Since that accident the nuclear industry has been desperately trying to convince the public that nuclear power is 'safe'. We are told that the radiation doses we receive as a result of nuclear power plants, nuclear reprocessing plants, nuclear waste dumps and nuclear weapons testing establishments, are at 'safe' levels, and that the background radiation these places produce is no more harmful than the natural radiation from granite in places like Cornwall.

What we are not told is that:

1 The 'safe' levels of radiation vary depending on where you live. For example, in Britain the 'safe' level of background radiation exposure for ordinary civilians is 5 millisieverts.* For workers in the nuclear industry, in Britain the 'safe' level is 50 millisieverts. In West Germany the safe level is 0.3 of a millisievert, while in the USA the safe level is 0.25 of a millisievert.

In other words, the 'safe' level for ordinary people in Britain is 20 times higher than the 'safe' level in the USA!

2 As far as the granite comparison goes, what the nuclear

---

* See glossary at back of book for explanation of this and other technical terms used.

industry does not tell you is that you would need to eat or breathe in 80,000 tonnes of granite to receive a lethal dose of radiation from it, while breathing in only a fraction of a gramme of plutonium will kill you. Such comparisons are bizarre, to say the least. After all, when did you last hear of anyone eating even an ounce of granite, let alone 80,000 tonnes?

The other area where the nuclear industry is fond of making comparisons is in their 'safety' record. They will proudly tell you that there has not been one death in the UK nuclear industry caused by a nuclear accident, and they will compare it with the accident rates in the gas and coal industries.

The trouble is that it is not possible to prove absolutely that the cancer that kills someone is a result of their working in (or living near) a nuclear plant. British Nuclear Fuels deny any such connections, but they are paying compensation to 13 families, with another 57 cases awaiting settlement, for cancer deaths of workers at their nuclear plants. It is also interesting to note that the nuclear industry used proudly to compare the accident figures of coal miners with uranium miners, until they discovered how many uranium miners developed cancers.

The real problem, though, is that of the long-term effect of a nuclear accident.

An accident at a coal mine or a coal-fired power station is dreadful and may kill people, but once that fire or explosion is under control, the worst is past. An accident at a nuclear plant could have lethal effects for hundreds, even thousands of years. Plutonium 239 for example, is lethal for 24,000 years, and uranium 233 for 162,000 years.

As you may have gathered, I am an opponent of nuclear power as well as of nuclear weapons, and in this book I intend to: **1** set out the anti-nuclear case; and **2** explain that there are alternatives to the nuclear industry that are safer, cheaper, and whose use will stop the nuclear industries (both weapons and energy) from destroying our planet.

It is my hope that when you finish reading this book you will agree with me, and want to work to save our planet from destruction.

At the back of the book I have included a list of organisations that you might like to contact. They will be able to give you more information, and can tell you what YOU can do to protect YOUR planet.

Jim Eldridge
September 1986

Part One
The Nuclear Industry

1 Nuclear Energy and Radiation

What is nuclear energy?

Nuclear energy is produced by splitting atoms.

Before we go any further it might be useful to answer some basic questions such as: How big is an atom? Is it easy to split one? Can I do it with a knife and fork? The answer to these questions are (a) tiny; (b) no; and (c) no.

To describe what an atom looks like, imagine our solar system, with the planets orbiting around the sun. The 'sun' is the nucleus of the atom, and the 'orbiting planets' are the electrons that orbit around the nucleus. (See Fig. 1)

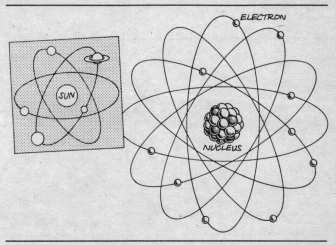

Fig. 1: Diagram of an atom (inset: our solar system)

The difference between our solar system and an atom is one of size: a tiny grain of sand, for example, would contain ten million million million atoms. Everything, whether it is liquid, solid or gas, is made up of these ultra-tiny atoms.

Splitting the Atom

To 'split the atom' means to split the nucleus at its centre. To picture this happening, imagine a snooker table with the red balls in their triangular shape, all ready for a game of snooker. This triangle of red balls is the nucleus, and the balls are a mixture of protons and neutrons, the two kinds of particles that make up the nucleus of an atom.

In the same way that this triangle of red balls splits up and goes in all directions when the white cue ball strikes them, so the protons and neutrons in an atom's nucleus split up when they are hit by a neutron which is 'fired' at the nucleus. (See Fig. 2)

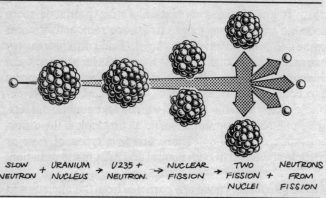

SLOW NEUTRON + URANIUM NUCLEUS → U235 + NEUTRON → NUCLEAR FISSION → TWO FISSION NUCLEI + NEUTRONS FROM FISSION

Fig. 2: How an atom splits

However, not all atoms act this way. Some, when they are split, merely sub-divide once, as if just one red ball has come out of the triangle. The nuclei of some atoms do not split at all. For an atom to split, you need an unstable atom, one whose nucleus is already in the process of changing. Such an atom is one that is giving off radiation naturally.

If you take such an atom and bombard it with neutrons you set off a chain reaction, which means that the atom's nucleus sub-divides, then sub-divides again, then sub-divides again, and keeps on sub-dividing, each sub-division giving off more and more energy. The bigger the chain reaction, the bigger the amount of energy produced. This process of splitting the nucleus of an atom is called nuclear (from the word 'nucleus') fission ('fission' derives from *fissio*, the Latin for 'to split'), and the energy given off by these unstable atoms splitting is called radiation.

How radiation affects living things

The problem with nuclear fission is that it depends on a chain reaction where an unstable atom produces more unstable atoms. When these unstable or 'radioactive' atoms come into contact with other atoms of living organisms, they can make those atoms unstable as well. As far as people are concerned, an unstable atom is like flu – whoever comes into contact with it is affected, and in human beings unstable atoms take the form of cancers and genetic defects.

To put this into perspective, let us first look at the different types of radiation. There are four types of rays or particles given off in both natural and man-made radiation:

Alpha particles These are particles from the nucleus and are not very penetrating. A sheet of paper or the outer layers of skin on the body can stop them.

Beta particles These are electrons and can penetrate only a few centimetres of flesh. They can be stopped by a sheet of metal or a block of wood.

Gamma rays These are very penetrating. Thick barriers of lead or concrete are needed to stop them. X-rays are like gamma rays and are produced when electrons bombard a metal target.

Neutrons These cause material they come into contact with to become radioactive.

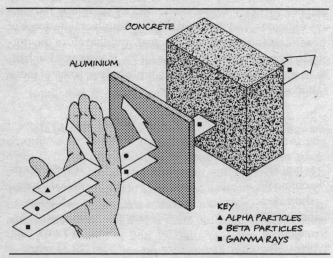

Fig. 3: The penetrating powers of different types of radiation

The problem is that as you can neither see radiation, smell it, or taste it, there is no way of knowing beforehand if you are about to be contaminated with it. Radiation can be absorbed into the body through breathing, eating or drinking.

When radiation comes into contact with a human cell, it damages it. The damaged cell may be killed outright by the body's defences, or the body may repair the damaged cell. If the cell survives and remains damaged it can reproduce inside the body. These damaged or deformed cells may later become a cancer. If the damaged cell is in the reproductive organs, then babies born to radiation-contaminated people may be stillborn or deformed, or the deformities (physical or mental) may be passed on to later generations.

The diagram of the human body (Fig. 4) shows which parts of the body are most affected by particular radioactive elements. The problem is in the length of time some of these elements remain radioactive, and therefore dangerous to living things. This danger level is measured in half-lives, and as you will see by the table below, some of these half-lives are very long indeed.

Isotope	Radiation Emitted	Half-Life
Uranium 233	Alpha	162,000 years
Plutonium 239	Alpha	24,000 years
Carbon 14	Beta	5,600 years
Caesium 1378	Beta, Gamma	30 years
Strontium 90	Beta	28 Years
Tritium (Hydrogen 3)	Beta	12.26 years
Krypton 85	Beta, Gamma	10 years
Cobalt 60	Beta, Gamma	5 years
Radium 266	Alpha	620 days
Ruthenium 106	Beta, Gamma	1 year
Zinc 65	Gamma	245 days
Polonium 210	Alpha	138 days
Sulphur 35	Beta	87 days

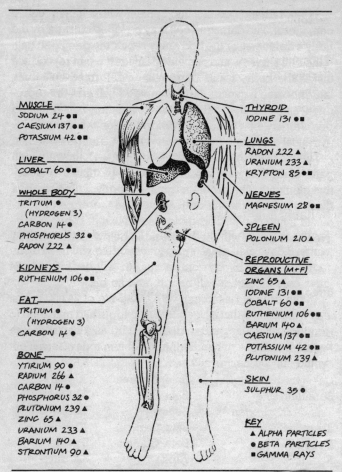

Fig. 4: How radiation affects human beings

Isotope	Radiation Emitted	Half-Life
Phosphorus 32	Beta	144 days
Barium 140	Gamma	13 days
Iodine 131	Beta, Gamma	8 days
Radon 222	Alpha	3.8 days
Ytirium 90	Beta, Gamma	64 hours
Magnesium 28	Beta, Gamma	21 hours
Sodium 24	Beta, Gamma	15 hours
Potassium 42	Beta, Gamma	12 hours

What is a half-life?

Saying that the half-life of iodine 131 is 8 days means that within 8 days, half of the radioactive iodine will have disintegrated. After 16 days only a quarter will be left, after 24 days only an eighth will be left, and so on.

In the case of plutonium, with a half-life of 24,000 years, this would mean that if Neanderthal people had managed to produce plutonium and then given it up as a bad idea, we would be unable to handle the plutonium today because it would still be dangerous.

2 Nuclear Power

A Nuclear Reactor

Splitting the atom to produce energy is one thing, harnessing that energy is quite another.

A nuclear plant produces energy in much the same way as a coal power plant: heat is produced which boils water which produces steam. This steam passes through turbines and these in turn drive generators which produce electricity. The difference is that in a nuclear power plant the heat is produced in a nuclear reactor, and the fuel that heats up the water or gas (depending on the type of reactor it is) is uranium. In fact it is not just ordinary uranium, but 'enriched' uranium.

Raw uranium contains two different kinds of uranium: uranium 235 and uranium 238. Uranium 238 is not a good fuel (although, as we will see in Chapter 4 'Nuclear Power – The Weapons Connection', it has a very important part to play). Uranium 235, on the other hand, is the more unstable of the two atoms and is ideal as a nuclear fuel. However, as only one per cent of uranium is uranium 235, raw uranium is 'treated' in order to extract the radioactive uranium 235. 'Enriched uranium' is the result of this extraction process and contains mostly uranium 235.

For those of you who are technically-minded and want to know what is the difference between uranium 235 and uranium 238:

> The nucleus of a **uranium 235 atom** has 92 protons and 143 neutrons: $92+143=235$.

The nucleus of a **uranium 238 atom** has 92 protons and 146 neutrons: 92+146=238.

The process of producing energy then follows the method outlined at the start of the previous chapter: the fuel elements of enriched uranium are bombarded with neutrons to split the atoms of the uranium 235. To slow the neutrons down and make sure they hit the atoms at the right speed for this splitting (fission) to take place, the fuel elements are separated from one another by a 'moderating' material (e.g. carbon). This material moderates the speed at which the neutrons travel.

There are also control rods in the reactor which can be slid in and out (by remote control) to absorb neutrons. These control rods regulate the neutrons, and therefore control the level of power at which the reactor works.

This whole arrangement of fuel elements, moderating material and control rods, is called the *core* of a nuclear reactor.

To convert the heat given off from this core into energy, gas or liquid known as 'coolant' is pumped through it. The coolant becomes hot, boils water, produces steam and so turns the turbines which drive the generators which produce electricity. (See Fig. 5)

The Problems

When described in these simple terms it all looks perfectly straightforward: uranium goes in at one end and electricity comes out at the other. Surely there is not that much difference between a nuclear power plant and a coal-fired power plant? After all, they both take in fossil fuel (coal in the case of one, uranium in the other) and produce electricity

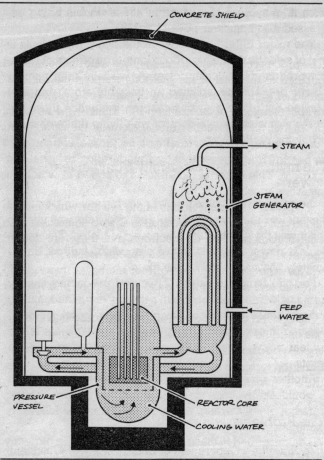

Fig. 5: How a Pressurised Water Reactor (PWR) works

from it. Why, then, are there all these doubts being expressed about nuclear power?

The major difference between the two power plants is that of safety, and the hazardous long-term effects of radiation produced by nuclear power. As we have seen, nuclear power depends on splitting an unstable atom, and as I pointed out in the previous chapter, it is when those unstable atoms come into contact with living things that our problems really begin.

The myth of cheap electricity from nuclear power

In the early days of nuclear power, the big selling-point was that nuclear energy was cheap. One of the slogans used to sell it was that it would be 'too cheap to meter'. This made nuclear power a very attractive proposition indeed. Unfortunately, it was wishful thinking.

For one thing the safety aspects of producing energy from nuclear fuel had not been properly examined, and so none of those early 'too cheap to meter' costs included building in safety measures to protect the workers in the nuclear power plants, or the people who lived around the plants.

Another problem was that the estimates of how much it would really cost to build a nuclear power station were hopelessly wrong, as can be seen from the following examples:

... in America a Department of Energy study of 47 nuclear plants found that 36 of them cost at least twice their original price, and 13 of them cost 4 times as much.

... a 1,100 megawatt reactor at Shoreham, Long Island, was originally expected to cost £160 million. In 1983 it was ten years behind schedule and the cost by then had gone up to £2,670 million, over 16 times the original estimate.

... when the first Magnox reactor was started up in 1962 at Berkeley in Great Britain, its electricity cost three times as much as electricity from the best coal-fired power stations.

... the AGR (Advanced Gas Cooled Reactor) at Dungeness in Great Britain recently became operational. Not only was it ten years late being finished, but its final cost was over £200 million, compared to the original estimate of £89 million.

> *'The capital cost of a nuclear power station, if spent instead on energy saving, would save three times more energy than the station would produce in its lifetime.'*
>
> Sir Martin Ryle, Astronomer Royal

Safety

More important than the fact that nuclear power is far more expensive than was originally thought, however, is the fact that nuclear power is more dangerous than any other form of man-made power. When discussing safety, the long-term effects of the radioactive elements involved in producing nuclear power must be borne in mind. Radiation-contamination is not like toothache, you do not just suffer it and then recover a few days later after visiting your friendly local dentist – radiation-contamination can last a long, long time.

As the nuclear industry grew, opinions began to be

expressed by some scientists, doctors and environmentalists that radiation from nuclear power plants was affecting the health of the people who worked in them, as well as those who lived nearby. It was suggested that these health risks were caused by radioactive gases leaking from the plants, and from nuclear waste being disposed of, often just pumped in liquid form straight into the sea right next to public beaches.

Fig. 6: How radiation enters the food chain

As a result of these doubts, the Government set up an enquiry to examine the health of children who lived near to the Sellafield nuclear plant at the nearby village of Seascale. Reports from Seascale had suggested that the rate of leukaemia in children there was very high indeed, much higher than the national average, and that the cause was radiation from the Sellafield plant.

The Government report, known as the Black Report, was

published in 1984 and the commission said that they could give a 'qualified reassurance to concerned people that they had found no evidence of a general health risk for children or adults living near Sellafield'. The report also stated that the incidence of leukaemia in Seascale is 'unusual but not unique', and that 'the suggestion that there is a relationship between an increased level of radioactivity and an above average experience of leukaemia is by no means proven.' In other words, the level of background radiation around the plant at Sellafield may be higher than most other places in Britain, but there was no connection with the high incidence of leukaemia.

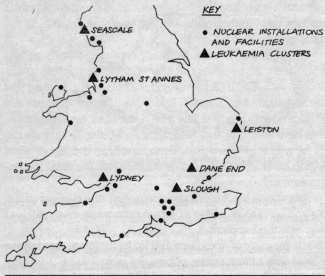

Fig. 7: The incidence of leukaemia in relation to nuclear installations

However, confidential copies of the evidence supplied to the Government Inquiry show that nearly *half* the cases of childhood cancer that occurred in Seascale and nearby villages were simply left out of the Black Report's figures. When the missing cases are included, Seascale has a leukaemia rate 24 times the national average.

It was this 'messing around with statistics' (a polite way of saying 'lying') by the pro-nuclear lobby that made those people who doubted the safety of nuclear power more suspicious than ever of the nuclear industry. In fact the more questions were asked, the more obvious it became that the nuclear industry was engaged in a cover-up at all levels about the true risks to health. This cover-up was particularly obvious in the summer of 1977, when a public inquiry was held into the Windscale (now Sellafield) nuclear plant.

At that inquiry it was originally reported that there had only been 28 accidents at the Windscale nuclear plant. Under oath, BNFL (British Nuclear Fuels Ltd) admitted that there had in fact been 177 accidents, most of them not disclosed to the public. The final figures that emerged were that between 1950 and 1977 there had been 194 accidents, 11 of which involved fires or explosions, and about 45 of which involved releases of plutonium into the working environment.

This was the accident record for just *one* nuclear plant. As you will see from Fig. 8, Great Britain has many nuclear plants dotted around it.

The majority of accidents and leaks that happen at these plants are never made public. In most cases it is only months after an accident that news leaks out. In 1985 there was a leak of radiation from the reactor at Dungeness, but it was not made public until 1986, and then it was only

Fig. 8: Britain's nuclear plants

admitted to by BNFL under pressure from a newspaper reporter. Also in 1986 a cloud of plutonium gas escaped from the reactor at Trawsfyndd nuclear power station.

These are just two recent examples. The point is that the safety record of the nuclear industry is not a good one. The reactors leak radiation, there are accidents involving radioactive materials, and yet the nuclear industry persists in telling us that nuclear power is 'the safest industry of all'. They have always said that the big accident, the major nuclear disaster, will 'never ever happen'.

That lie was exploded on 26 April 1986 when the nuclear reactor at Chernobyl in Russia blew up.

3 Chernobyl

We should have learned our lesson about expert opinion on safety matters in 1912, after the experts said that the *Titanic* was unsinkable. An iceberg proved them wrong.

A major accident at a nuclear power station, we were told, could never happen. This was because the possibility of a major accident was 'one in every 10,000 years'. However, this figure was for *one* nuclear reactor. As there are 211 nuclear reactors operating in Europe (see Fig. 9) this alters the odds considerably, and the theoretical risk of a

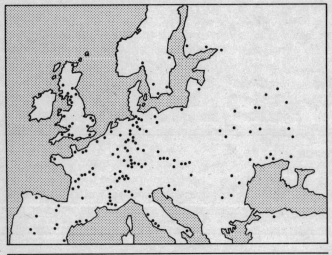

Fig. 9: Nuclear power reactors in Europe

major nuclear accident becomes 1 in (10,000 divided by 211), which equals one major accident every 47.4 years.

However even these statistics do not take human error into account. When you consider that nuclear power plants are designed, built and operated by human beings, it means that human error must be taken into consideration. Even the Chairman of the CEGB (The Central Electricity Generating Board), Lord Marshall, has admitted that if human error is added into the calculation, it would throw all the 'error-probability' statistics (a long winded way of saying 'when one blows up') out by 80%.

Which brings us to the fact that one did blow up at Chernobyl on 26 April 1986.

Although there is still much confusion over what happened at Chernobyl to cause the accident, it is now generally assumed that human error was to blame. In the early hours of the morning of Saturday 26 April, the Number 4 reactor at the Chernobyl power station was operating at only 7% power. In fact, it was just ticking over while maintenance work was being carried out. Although heat was still being produced in the core, the chain reaction had been shut down. Suddenly the power went from 7% to 50% in ten seconds. This sudden surge of power sent temperatures inside the reactor soaring. It is believed this sudden surge of power was caused by the operator experimenting with the control rods. This was the human error.

Scientists suggest that what happened then was this: the zirconium metal (which was used as a cladding for the uranium fuel) reacted with steam at this sudden high temperature (1000° centigrade) and produced hydrogen. The pressure caused by the steam and the hydrogen blew the top off the reactor core.

The hydrogen escaped into the containment area.

Hydrogen burns explosively when it comes into contact with oxygen, and when it came into contact with the oxygen in the containment area it did just that, blowing a hole in the roof of the reactor hall. The explosion also toppled the reactor's 200 tonne crane (used for refuelling the reactor) down onto the core.

In addition, broken pressure tubes were no longer providing essential coolant to the core. This caused further overheating, which caused further reactions between the steam and the radioactive elements in the core.

By the time the fire service arrived, the fire from the No. 4 reactor had nearly reached the No. 3 reactor, and there was now danger of the fire spreading across the whole Chernobyl complex.

The Russian fire fighters soon realised that they could not put the fire out easily, so they decided to contain it in the No. 4 reactor and so stop it spreading. They did this by spraying the roof of the No. 3 reactor with water to cool it. In order to spray the No. 3 reactor they had to fight the fire from one of the station's highest roofs. Because of the intense heat, the bitumen on that roof melted, and the firemen found their boots sticking to it.

However, their containing action succeeded, and although the fire was still burning 36 hours after it started, it was contained in reactor No. 4. All the firemen were contaminated with radiation from the fire, and many died shortly afterwards.

Actually extinguishing the fire meant calling in the Soviet Air Force. The Air Force flew 279 helicopter sorties over the burning reactor and dropped a total of 5,000 tonnes of a smothering mixture onto the fire. This smothering mixture was made up of sand, clay, lead, dolomite, and boron. The boron was to absorb neutrons and so prevent the chain

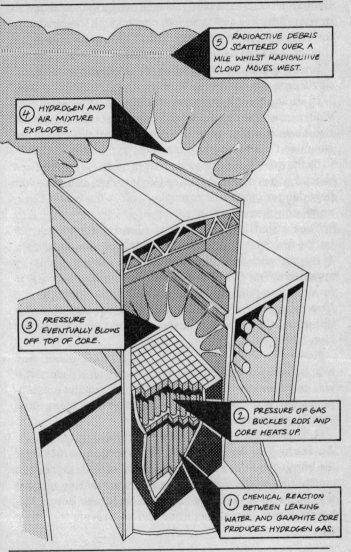

Fig. 10: Chernobyl – how it happened

reaction beginning again; the lead was to shield against gamma radiation; and the sand, clay and dolomite were to put the fire out by smothering it, and also to absorb as much of the fission products as they could.

Since the accident the design of the Russian reactor has been condemned by many Western nuclear experts, including Lord Marshall, the Chairman of the CEGB, in an effort to allay fears of a similar accident happening in the West. Two things are worth bearing in mind when listening to these criticisms:

1 In 1976 this same Lord Marshall, at an International Symposium in Zurich, praised this same Russian reactor design as highly reliable.

2 The design of the Chernobyl reactor actually helped to prevent a melt-down. This is because in the Chernobyl reactor the fuel is contained in different channels. In a Western-designed PWR (the sort it is proposed to build at Sizewell) all the fuel is in the same pressure vessel, and if there is an accident it can accumulate at the bottom of this vessel in a large molten mass. It is this large molten mass of fuel that is likely to lead to a melt-down, the 'China Syndrome'.

Despite the firemen's efforts, a radiation cloud had already escaped from No. 4 reactor and was now in the outside atmosphere. The rest of the damage would be done by the wind, which carried the radiation from Chernobyl more than 1,600 km.

Most of the plutonium and uranium elements stayed inside the Chernobyl complex, or fell in the local area. Other radioactive elements (e.g. caesium 137, iodine 131, strontium 90, ruthenium 103) were in the cloud and eventually they contaminated nearly every country in Europe. (See Fig. 11)

The contamination and the reaction to the contamination in the rest of Europe (outside the Soviet Union) was as follows:

In Switzerland, radiation levels were ten times higher than the normal. Drinking rainwater was banned and pregnant women and children were warned not to drink milk. Farmers in Berne were advised against feeding green fodder to their livestock.

In Sicily and Sardinia the sales of milk and leafy vegetables were restricted.

In West Germany some areas banned the harvesting of certain leaf vegetables. Parents were told to bathe their children if they played outside. At Kehl, on one side of the Rhine bridge, children were forbidden to play on the grass. Meanwhile, on the French side of that same bridge, no action was taken, which raises the question: Why not?

The answer lies in the fact that France is totally committed to nuclear power and has spent years telling the French people how safe and cheap it is. At first the French Government refused to admit that there was any radiation from the Chernobyl disaster. Eventually, however, they admitted that background levels in France had risen up to 400 times higher than normal background radiation, but these incredibly high radiation readings had been concealed from the public.

In Britain a similar reassuring 'no need to be alarmed, it's all perfectly safe' approach was taken by the Government for much the same reason – the Government was determined to press ahead with its nuclear power programme. It wanted to build a new PWR (Pressured Water Reactor) at Sizewell in Suffolk; and once that one was established it wanted to build another eleven around the country. To

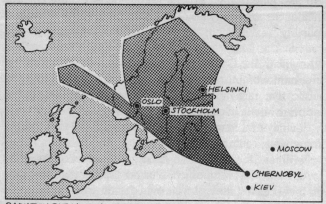

RADIATION PATH DURING FIRST WEEK

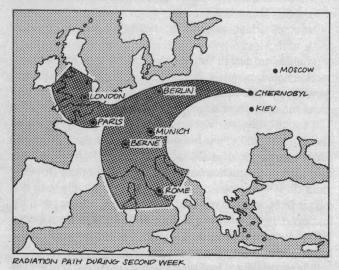

RADIATION PATH DURING SECOND WEEK

Fig. 11: The path of radioactive cloud from Chernobyl

tell the public that Britain might be affected by the nuclear disaster at Chernobyl would be to alarm them and possibly make them resist all plans for new power stations.

However, the Government had to say something, so the Environment Minister at that time, Kenneth Baker, told the public that the radioactive cloud would not present a hazard to health. Government figures showed (he said) that the levels of radioactive iodine 131 in milk only measured 60 bequerels a litre. Unfortunately for the Minister and the Government, the National Radiological Protection Board (a pro-nuclear organisation) published figures that said that the levels of iodine 131 in milk in Britain were actually as high as 390 bequerels a litre; a considerable increase on the official Government figures.

Eventually, the Government was forced to ban rainwater for drinking; and advised against the consumption of leafy vegetables in certain areas, and lambs that had fed on contaminated pastures. By that time, a month after the accident, people had already drunk radioactive water and eaten radioactive meat and vegetables.

What the Chernobyl disaster also showed is that the British government's emergency precautions against a major nuclear accident are more suited to a Daffy Duck cartoon than a real nuclear emergency. None of the Government departments were able to cope effectively with the fallout from a nuclear accident, 1,600 km away, yet they persisted in telling the public that there was no need to worry, and that they could deal with this sort of accident if it happened at a British nuclear reactor.

Before you start feeling safe just because the British Government tells you that you are, it is worth looking at what the Government's emergency plans are in the event of a major accident at a nuclear reactor in the United Kingdom. The best example is their most recent contingency plans in the event of a major accident at the Sizewell reactor in Suffolk:

1. Evacuation of people living within 2.4 km of the nuclear power station.

2. A ban on sales of milk up to 40 km from a damaged reactor.

3. Injured and contaminated people to be taken to the nearby Ipswich hospital for treatment.

Before you start thinking 'Good, they do know what they're doing', bear in mind that the immediate things wrong with these contingency plans are:

1. This '2.4 km' figure is obviously absurd. It assumes that any wind blowing would abruptly stop when it had reached that distance from the damaged reactor. In reality (as with Chernobyl) a major nuclear accident would affect a far greater area. In Sweden, the official evacuation zone is between 40 and 80 km. At Chernobyl, the Soviet authorities had to evacuate everyone living within 30 km of the reactor.

2. This is based on a Government Ministry nuclear emergency plan that says 'It is unlikely that milk would be affected as far away as 40 km. In reality, after the Chernobyl disaster milk was banned as far away as 1,200 km.

3. The reaction from Ipswich hospital to this was that they could only deal with four or five victims at a time,

so this Government 'emergency plan' is based on their hope that a major nuclear disaster would only contaminate people in bunches of five, and then no more than five at a time.

One big question remains: could a major disaster like the one that happened at Chernobyl happen in Britain? The short answer is: Yes.

> *'A major commitment to fission power and the plutonium economy should be postponed as long as possible because of its grave potential implications to mankind.'*
>
> *A British Government report, 1976*

After the Chernobyl disaster the nuclear industry provided 'experts' in white coats to appear on television and reassure the public that a Chernobyl disaster could not happen here because our nuclear reactors are different from the Russian nuclear reactors. For one thing, they said, the reactor at Chernobyl did not have a secondary containment shield around it to stop the radiation cloud escaping. True, it did not. However, what they forgot to add, was that none of Britain's Magnox reactors have secondary containment shields around them either.

Something else they forgot to mention was that there had already been two major accidents to nuclear reactors in the West: one at Windscale in Britian in 1957, and one at Three Mile Island in the USA in 1979.

The Windscale accident happened on 10 October 1957 when fire broke out inside the reactor. Uranium fuel and cladding was on fire and melting. The fire was put out by the following day, but by that time a cloud of radioactive dust had escaped. A report from the National Radiological Protection Board (who, as I pointed out earlier, are a pro-nuclear organisation and therefore unlikely to be exaggerating) said it was likely that 260 cancers were caused by this fire. In addition, all the milk from cows grazing within 200 miles of the Windscale plant had to be poured away.

At Three Mile Island in the USA, the accident happened when coolant from the reactor evaporated and there was no more to replace it. The uranium fuel rods were exposed, the fuel alloy cladding began to melt and react with the steam, and the uranium began to melt.

Fortunately, the reactor operators prevented the core from melting into the earth, but only just. A report from America's Nuclear Regulatory Commission stated that the nuclear reactor at Three Mile Island had only been one hour away from a core meltdown.

A core meltdown would occur if all the reactor coolant leaked or boiled off from a reactor core. If that happened the reactor's uranium or plutonium fuel would melt into one large lump, a lump that would produce such heat that it would be beyond control. The lump would then simply melt through solid concrete and rock and sink into the ground. As it burned into the earth it would explode and send out highly radioactive waste into the atmosphere, devastating the area around the reactor.

These are two examples of nuclear power station accidents that have happened in the West. The chances of a major accident happening in the future here in Britain increase every day because of the age of our nuclear reactors. Most of our Magnox nuclear reactors were built in the 1960s. Since 1979, cracks have been found in the welds of the cooling systems of these Magnox reactors, particularly at Bradwell, Dungeness, Hinkley Point, and Sizewell. If one of these cracks led to the cooling system fracturing it would mean that the core could overheat, with disastrous results similar to the Chernobyl accident. There are also cracks in the roof beam supports at Dungeness, which could bring the concrete roof crashing down on the reactor, which again could lead to a Chernobyl-type disaster.

In view of the Chernobyl disaster, the near meltdown at Three Mile Island, the various accidents at Windscale (Sellafield) and Britain's other nuclear reactors, and the possibility of a major accident at a nuclear power station in the future, why do Governments continue to follow a path that promotes nuclear power? Surely it is obvious to anyone that nuclear power is dangerous, certainly too dangerous to take chances. One major accident could destroy a whole continent for generations to come.

Why, then, is it still the favourite form of power for so many governments, Britain included? It certainly is not because of its cheapness, and it certainly is not because of its safety record. As we shall see in the next chapter, the answer lies in the connection between nuclear power and nuclear weapons.

4 Nuclear Power – The Weapons Connection

The story of nuclear power starts in the Second World War with the race between the Western Allies (particularly the USA and Britain) to produce a nuclear bomb before Hitler's Nazi Germany.

The US Government, worried by reports that Nazi Germany was conducting research into atomic science, set up the Manhattan Project in May 1942. The project was led by Robert Oppenheim, and its aim was to turn the theory of atomic physics (nuclear fission, chain reactions, etc.) into practice in the form of a bomb that would win the war.

In December 1942 a group of Manhattan Project scientists, led by Enrico Fermi, created the first man-made nuclear chain reaction. They did this in the world's first nuclear reactor, constructed in a squash court at the University of Chicago, and made out of 350 tonnes of graphite, 50 tonnes of uranium oxide, and 6 tonnes of uranium. Over the next three years the Manhattan Project carried out its bomb research at Los Alamos in New Mexico. At the same time two engineering plants, one at Oak Ridge, Tennessee, and one at Hanford, Washington, were producing the Uranium 235 necessary for the weapons.

By 1945 the first atomic bomb was ready. It was codenamed 'Trinity', and it was exploded at Alamagordo, New Mexico, on 16 July 1945. Made with plutonium as its explosive, its explosion was equivalent to 18,000 tonnes of TNT.

The following month (August 1945) America used its new

nuclear weapons for the first time against people – at Hiroshima and Nagasaki. The Hiroshima bomb was a uranium bomb, while that dropped on Nagasaki was a plutonium bomb. (We shall look at the actual explosions and their effects in more detail in Chapter 5).

Although some British Government scientists had worked on the Manhattan Project, by the end of the war in 1945 the US Government had become worried about security, and had decided that they did not want the British Government to know the secrets of American nuclear weapons research. The British Government decided to produce a nuclear bomb of its own.

The British scientists who had worked on the Manhattan Project had already decided from their work in America that plutonium made a better bomb than uranium, so in December 1945 the British Government, acting on their advice, decided to build an atomic pile that would give Britain enough plutonium to make 15 bombs a year. This atomic pile was to be at Windscale in Cumbria. Next to the pile a reprocessing plant was built. Its purpose was to reprocess nuclear fuel and extract the plutonium necessary for weapons. By 1952 Britain's first nuclear bomb was ready, and in September of that year it was tested in the Montebello Islands off the north-west coast of Australia. The bomb worked.

Now that Britain had the capability to produce a nuclear bomb, it wanted more bombs, and to make more bombs it needed more plutonium. Four reactors were built at Calder Hall, and another four at Chapelcross. Their aim was to produce plutonium for Britain's nuclear weapons.

To explain how reactors produce plutonium it is necessary to go back to Chapter One and the explanation of how the atom is split. When an atom of uranium 235 is bombarded

> *'Our consideration has led us to the view that we should not rely for energy supply on a process that produces such a hazardous substance as plutonium unless there is no reasonable alternative.'*
>
> *Royal Commission on Environmental Pollution, Sixth Report, (September 1976)*

with neutrons it splits and sets off a chain reaction. During this chain reaction, neutrons bombard the atoms of uranium 238. Uranium 238 was previously considered useless as a fissile material (one good for nuclear fission), but if its nucleus has another neutron added to it, then it becomes uranium 239 (an unstable atom), and uranium 239 changes into plutonium 239.

Note: these atoms are all named after the planets – uranium after Uranus, neptunium after Neptune, which is the next farthest planet out in the solar system, and plutonium after Pluto, which is the *next* farthest one.

During the process of converting uranium to plutonium for bombs it was discovered that an excess of energy was given off, and that this excess of energy could be harnessed to produce electricity. It was in this way that Britain's civil nuclear power programme began.

The Calder Hall reactor was opened by Queen Elizabeth II on 17 October 1956, and it was proudly announced that Calder Hall would be producing electricity for the National Grid (the British electricity system). What was not publicly announced was that the real purpose of the reactor at Calder Hall was to produce plutonium for bombs. Electricity for the National Grid was an afterthought.

For the bomb-makers the unfortunate fact about plutonium is that it does not exist naturally, it is man-made. And as the demand for weapons increased, so did the demand to make more plutonium.

One way to meet a demand like this is to build more and more nuclear reactors, something governments across the world keep trying to do. However, as doubts about the safety of the nuclear industry increase, so it takes longer and longer for governments to overcome opposition and to build nuclear reactors. In the meantime plutonium has to come from another source, and this is where the reprocessing of nuclear waste comes into the picture.

The phrase 'nuclear waste' covers a multitude of things, as we shall see in Chapter 6. Among others it includes spent nuclear fuel from nuclear power stations and from the reactors of nuclear submarines. This spent nuclear fuel contains plutonium. To extract this plutonium it needs to be reprocessed.

To understand reprocessing, imagine a load of mucky sludge that contains a few gold and silver coins. The smaller gold coins are the plutonium, and the larger silver coins are the uranium. To find the coins, the mucky sludge is poured into a huge vat of water. This separates the sludge and turns it almost back into liquid. The whole mess is sieved through two large sieves, one with fine mesh (for the gold coins), and one with larger mesh (for the silver coins). The rest of the watery mess is collected in a large vat.

So now the plutonium (gold coins) and the uranium (silver coins) have been extracted from the sludge. But there is a lot more sludge than before because of the extra water. All of this sludge is now contaminated, which is why reprocessing nuclear waste produces more nuclear waste than there was in the first place.

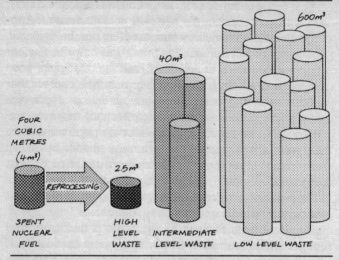

Fig. 12: How reprocessing nuclear waste produces more nuclear waste

The official reason for reprocessing nuclear waste given by the nuclear industry (organisations like British Nuclear Fuels Ltd and the Central Electricity Generating Board) is that it is necessary in order to extract uranium, which is needed as the fuel for nuclear power stations. This is an odd reason, because it is actually cheaper to mine uranium and to buy it in its raw state and then enrich it, than it is to produce it by reprocessing spent fuel. It is also easier to produce it in large quantities by mining than it is by reprocessing. So what is the real reason for reprocessing nuclear waste?

There is only one reason for it: to produce plutonium for nuclear weapons.

That this is being done is denied by the nuclear industry, and with good reason. To use civil nuclear waste to produce plutonium for military purposes (i.e. bombs) is illegal. However, the fact remains that Britain has only one nuclear reprocessing plant, at Sellafield, and into that plant goes nuclear waste from both the military and the civil nuclear programmes, and out of it comes plutonium of different grades, including weapons grade plutonium. It was admitted by the CEGB in 1986 during a television interview, when they were asked to explain why there seemed to be quite a few tonnes of plutonium missing from the civil nuclear programme after reprocessing, that 'maybe' some weapons grade plutonium had come from civil nuclear waste, but that this had not been 'deliberate'.

The fact remains that with an increase in the numbers of nuclear weapons more and more plutonium is needed, and nuclear power and nuclear waste reprocessing is the only source. And just how destructive those nuclear weapons are, and exactly how many of them there are threatening our world, we shall see in the next chapter.

5 Nuclear Weapons

Hiroshima and Nagasaki

On 6 August 1945, a uranium bomb was dropped on Hiroshima in Japan. The bomb exploded 600 metres above the centre of the city. The fireball that resulted was 5 km across and had a temperature at its centre of about 56 million degrees Celsius. Over 10 km square of the city were destroyed. About 88,000 buildings (houses, blocks of flats, schools, offices, police stations, etc) were destroyed. Fires raged for two days. Some people near the centre of the explosion simply evaporated. Others were burnt to death. Some survivors had all their skin burnt off. 70,000 people were killed instantly, and 76,000 were injured.

On 9 August, three days later, a plutonium bomb was dropped on Nagasaki. The Nagasaki bomb exploded about 500 metres above the city. 38,000 people were killed instantly and 21,000 were injured.

By the end of October 1945 the death toll for both cities had risen to 130,000 at Hiroshima, and 65,000 at Nagasaki. In March 1973, 28 years after the bombs dropped, there were still 366,523 people registered as sufferers from the effects of the bombs. (These sufferers are known in Japan as 'Hibakusha' – 'survivors of the bomb'.)

Megaton bombs

The nuclear bombs dropped at Hiroshima and Nagasaki had the power of about 20 kilotons each. Nowadays we measure the power of most nuclear weapons in megatons (1 megaton = 1,000 kiloton). A one-megaton bomb would be 500 times more powerful than the nuclear bomb dropped on Hiroshima.

The effects of a nuclear explosion

The three main features of a nuclear explosion are: **1** the blast itself; **2** intense heat; and **3** radiation.

The blast The force of the blast will collapse buildings, crushing the people in them, hurl debris and glass around, crushing people and cutting them so that they bleed to death. The blast can also rupture internal organs such as the lungs, the heart, etc.

Heat The explosion's fireball can start individual fires, which can join together to make one huge fire. This becomes a fire-storm, in which very strong winds are sucked in by the fire. This can lead to temperatures of 1000 degrees Centigrade. At this temperature, glass and metals melt and people are incinerated.

Radiation For several hours after a one-megaton explosion, unprotected people as far as 100 miles downwind of the blast may receive lethal doses of radiation from the fall-out. Because of this it is impossible to say how many people were *actually* killed by the nuclear bombs at Hiroshima and Nagasaki.

The effects of radiation on people following a nuclear explosion can be seen from the following table:

Dose (in rads)	Symptoms
Below 100	Possibly some nausea and vomiting. *Deaths average*: 0

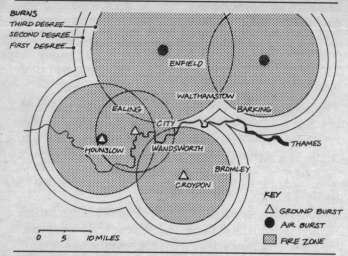

Fig. 13: Fire and burns after a nuclear attack on London

Dose (in rads)	Symptoms
100–200	Nausea and vomiting within 3–6 hours of receiving dose and lasting less than one day, followed by no symptoms for 2 weeks. Recurrence of symptoms for another 4 weeks. The number of white blood cells reduced. *Deaths (average):* 0
200–600	Nausea and vomiting lasting 1–2 days. No symptoms for 1–4 weeks, followed by a recurrence of symptoms for up to 8 weeks. Diarrhoea, severe reduction of white blood cells, blood blisters on skin, bleeding and

Dose (in rads)	Symptoms
	infection. Above 300 rads: loss of hair. *Deaths (average)*: 0–98% in 2–12 weeks from internal bleeding or infection.
600–1,000	Nausea and vomiting starting within half an hour of receiving the dose of radiation and lasting 2 days. No symptoms for 5–10 days, then the same symptoms as for 200–600 rads for 1–4 weeks. *Deaths (average)*: 98–100% from internal bleeding or infection.
1,000–5,000	Nausea and vomiting starting within half an hour of receiving the dose and lasting less than a day. No symptoms for about seven days, then diarrhoea, fever, and disturbed salt balance in the blood for 2–14 days. *Deaths (average)*: 100% within 14 days from collapse of circulation.
over 5,000	Nausea and vomiting immediately, followed by convulsions, loss of control of movement and lethargy. *Deaths (average)*: 100% in 48 hours from failure of breathing or brain damage.

How big an area would a one-megaton nuclear bomb affect?

The best way to answer this is to examine Fig. 14.

Civil Defence

Is there a way to protect yourself from a nuclear attack?

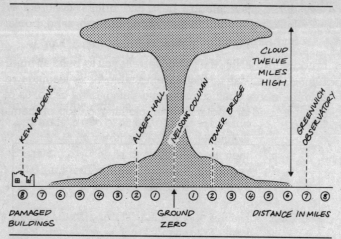

Fig. 14: The effect of a one-megaton nuclear bomb exploded at 2,000 metres above Trafalgar Square

Opinions vary on this: the Government say Yes; opponents of nuclear weapons say No.

The Government's official line on what to do in the event of a nuclear attack on Britain is contained in the pamphlet 'Protect and Survive'. It tells that you must make one of the rooms in your house a fallout room to protect you against radiation. It then goes on to say:

> 'Even the safest room in your house is not safe enough, however. You will need to block up the windows in your room, and any other openings, and to make the outside walls thicker, and also thicken the floor above you to provide the strongest possible protection against the penetration of radiation. Thick,

dense materials are best, and bricks, concrete or building blocks, timber, boxes of earth, sand, books and furniture might all be used.'

In addition to all this the pamphlet tells you to 'build an inner refuge', which can be made from a table, or by leaning two or three doors against a wall and piling 'bags or boxes of earth or sand or books or even clothing' on top of them.

Fig. 15: Building an inner refuge

According to the Government you should stay in this inner refuge for 48 hours. After that you should stay in your 'fallout room' for another 14 days. You should, of course, have stocked up with enough water and food for this length of time, because all the water and food outside your fallout room will be contaminated with radiation.

All this sounds very reassuring (as it is meant to). It suggests that all you have to do in the event of a nuclear war is to build an incredibly strong fallout shelter as soon as you know a nuclear bomb is on its way, get inside it with your family and all your supplies, and come out 16 days later when everything will be all right again.

In practice it will not be that simple. In 1980, a BBC film team made an indoor fallout shelter following the Government's instructions for the documentary programme 'Panorama'. Their construction took 100 bags and strength enough to lift a tonne of earth. In 1980 the Government was allowing 2 to 3 weeks as the warning period before a nuclear attack. In fact the warning period might be measured in hours, or even minutes. You might be just on your way out of your front door to look for the concrete blocks when the bomb actually strikes.

Also, far from any survivors emerging into a normal (if a bit charred) world after a nuclear attack, the landscape would be devastated. Even the Government admitted in a Home Office circular, published in 1976 and titled 'Environmental Health in War', that 'water would not flow from the tap or into the sewerage system. Electricity would be cut off, refuse collections would cease, large numbers of casualties would lie where they died. In such conditions disease would spread rapidly.'

Other Government circulars paint an equally gloomy picture:

On food 'Food would be scarce. Peacetime systems of food processing and distribution would cease to function. No arrangements could ensure that every surviving household would have 14 days supply of food after attack.' (Home Office circular, January 1979)

- **On water** 'Any widespread nuclear attack would quickly disrupt the distribution system for domestic and industrial water and much of the sewerage system.' (Home Office circular, June 1976)

- **On energy** 'After a nuclear attack all energy production and supply would soon cease.' (Home Office circular, April 1976).

EMP

There would also be the problem of EMP. What is EMP? The letters stand for Electro Magnetic Pulse, and this would put a halt to the plans that the Government have for keeping everyone informed about a nuclear attack. An Electro Magnetic Pulse occurs during a nuclear explosion. A large nuclear explosion would produce an EMP capable of destroying radio and telephone communications over a whole continent.

Target Areas

So where in Britain would be a likely target in the event of a nuclear war? As most of our Western politicians keep telling us that the Russians would be our enemy in a nuclear conflict, it is interesting to look at a Soviet map of Britain that points out places such as military air bases, naval bases, communication centres, etc. (See Fig. 16)

To this 1980 map need to be added the two American cruise missile bases in Britain at Greenham Common and Molesworth. As cruise missiles can be moved around up to 160km from their base, it has been worked out by British and American military chiefs that to eliminate these missiles, as well as other important targets, the Russians would have to drop 600 one-megaton bombs on Britain. And if 600

Fig. 16: A Soviet map detailing military air bases, naval bases, communication centres etc

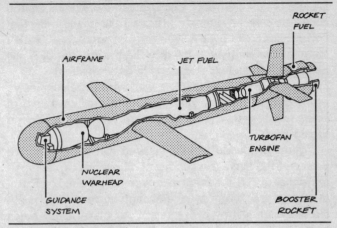

Fig. 17: A nuclear missile

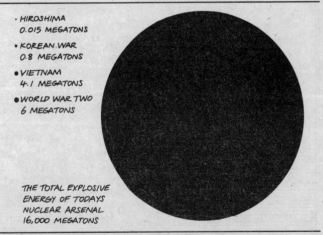

Fig. 18: The destructive force of today's nuclear weapons

nuclear bombs seem like a large number of bombs for one country to own, let alone drop, it is worth remembering that Soviet Russia has over 8,000 nuclear warheads all within range of Britain. (Just to balance the picture, the USA has nearly 11,000 nuclear warheads in its arsenal.)

A single *one* of these nuclear warheads has an explosive power greater than the total of *all* the two million tonnes of bombs that were dropped on Germany in the Second World War.

Figures produced in 1982 showed that the destructive power of the nuclear weapons of all nations represented *ten tonnes* of TNT for every man, woman and child on Earth. In other words we now have enough nuclear weapons in the world to destroy everyone on our planet many times over. (See Fig. 18)

> *'In the event of a nuclear war there will be no chances, there will be no survivors – all will be obliterated . . . In all sincerity as a military man, I can see no use for nuclear weapons which would not end in escalation, with consequences no one can conceive. Their existence only adds to our perils because of the illusions they have generated.'*
>
> Field Marshall Lord Mountbatten

An accidental war?

But, we are told, no one really wants a nuclear war; these nuclear weapons are only deterrents, to stop the other side from starting one. No one would really start a nuclear war.

Even if we accept that (and I for one do not. After all, someone is always starting a war somewhere), what if a nuclear war should start by accident? Our weapons and radar systems are now computer-controlled. It would take only one mistake. And what about our old enemy, 'human error'?

The following were times when a nuclear war nearly started 'by accident':

5 October 1960 Radar malfunction in ballistic missile early-warning system at Thule, Greenland, falsely warns North American Air Defense (NORAD) headquarters of 'massive' Soviet missile approaching USA.

20 February 1971 Operator at NORAD headquarters accidentally transmits emergency message ordering all US broadcasting stations off the air by order of the President. For forty minutes the operator cannot find the code with which to cancel the message.

9 November 1979 False warning of limited missile attack sent from NORAD headquarters to nationwide defence command centres. Interceptor planes scrambled and missile bases put on alert.

3 and 6 June 1980 False alarm indicating Soviet missile attack registered by NORAD computers at Colorado Springs. One hundred nuclear-armed B-52s alerted for take-off.

In addition to these there were also accidents involving nuclear weapons that could have been interpreted as a first strike by the other side:

2 February 1958 B–47 bomber in mid-air collision near Hunter Air Force Base, Georgia, accidentally jettisons part of a nuclear weapon.

4 June 1962 Thor ICBM (Inter-Continental Ballistic Missile) fails in high-altitude thermo-nuclear weapon tests. One-megaton warhead destroyed over Johnston Island Pacific Test Range.

20 June 1962 Second Thor failure. Missile and warhead destroyed at altitude of 320 km.

17 January 1966 B–52 carrying four H-bombs collides in mid-air with a KC–135 tanker near Palomares, Spain, and crashes.

21 January 1968 B-52 crashes near Thule, Greenland. Four H-bombs lost.

15 October 1969 B-52 carrying two nuclear weapons collides with KC-135 near Glen Bean, Kentucky.

23 October 1975 Canister containing 20-kiloton bomb falls down test shaft at Nevada Nuclear Test Site.

18 September 1980 Rocket fuel explosion due to 'human error' blows a Titan I warhead from its silo near Damascus, Arkansas.

These are all examples from the Western side. The Soviets are even more secretive about their nuclear activities than the West, but their nuclear weapon accidents include:

1968 A G-class ballistic-missile submarine sinks after an on-board explosion in the Pacific.

1970 An N-class nuclear submarine sinks in the Eastern Atlantic.

September 1974 A Kashin-class guided-missile destroyer carrying nuclear weapons explodes and sinks in the Black Sea.

August 1976 A Soviet nuclear-powered cruise-missile submarine collides with the American frigate *Vago* in the Ionian Sea.

October 1986 A Soviet submarine carrying nuclear missiles catches fire and then sinks off the coast of America. Nuclear weapons now lying on the seabed and liable to corrosion.

Nuclear Winter

No book that deals with nuclear weapons would be complete without at least touching on the subject of the nuclear winter.

On page 56, in the section on Civil Defence, I quoted from the Government's 'Protect and Survive' guidelines. These suggest that there would be a semblance of normality about three weeks after a nuclear attack. This three weeks time-scale was based on the idea of the radioactive fallout decaying to safe levels after that length of time (although even this is disputed by scientists).

However, in 1983 separate groups of scientists in both the East and the West came to a different conclusion about the after effects of nuclear war. According to them, the millions of tonnes of debris thrown into the air by the explosions and the fires would cause a thick layer of cloud to blot out the heat and light from the sun for many months. If this happened there would be almost total darkness. Temperatures would go down to 20 or 30 degrees Celcius below freezing. People and animals would die from starvation and cold. Nothing would grow. Water would be frozen. This would be in addition to the initial blast, heat and radiation effects of the nuclear explosions. In fact, some scientists went as far as to say that, with this theory, there might eventually be no survivors in the northern hemisphere, which is not a comforting thought.

Weapons Proliferation

The question is, how has all this come about? Why do we have these huge nuclear arsenals that can destroy all life on our planet many times over? And why do governments keep adding to their stockpiles with even more destructive weapons?

The answer lies in the equation: fear and power.

Ever since man first became a fighting animal, fighting for territory, he has worked to get bigger and better weapons than his enemies (these include anyone who might turn out to be an enemy at some distant time in the future). If a Stone Age man of one tribe had invented a spear that could travel for 50 metres and kill one person, as soon as a rival tribe got to hear about it they would get their weapons expert to produce a spear that could travel for 75 metres. The first tribe would then try and come up with a spear that could travel 120 metres, and so on. This is what is known as 'Weapons Proliferation' or 'Weapons Escalation'.

At the time of writing our two 'tribes' are assumed to be: Tribe A – the Western Allies (also known as NATO), which include Great Britain, the USA, Canada, France, West Germany, Holland, Belgium, Denmark, Italy, Luxembourg, Norway, Greece and Turkey; Tribe A also includes sympathetic countries such as Australia. Tribe B – the Eastern bloc countries (also known as the Warsaw Pact, or 'the ones behind the Iron Curtain'), which include Russia, Poland, Romania, Bulgaria, East Germany, Czechoslovakia and Hungary.

(In addition, there are a few other smaller 'tribes' who are interested in owning their own nuclear weapons, and could possibly provoke a nuclear war. These include countries like Libya and Iran.)

Why did this East versus West conflict arise in the first place? The answer lies back in 1945 at the end of the Second World War. At that time the USA and USSR were allies, along with Britain, France, Australia etc., having just fought together to defeat Hitler's Nazi Germany, Mussolini's Italy and Japan. However, the USA and USSR were allies who did not trust each other because they had opposing philosophies of government. The USSR had a communist government and saw capitalism as the real enemy. The USA opted for free enterprise and large profits and saw communism as the real enemy.

Both the USA and USSR felt threatened by the other. The Americans in particular were convinced that the USSR was planning to take over the world for communism. Certainly, after 1945 many countries seemed to change their form of government to communism, either as a result of election, revolution or invasion:

1946 Bulgaria becomes a communist people's republic.
King Zog of Albania deposed. Albania declared a republic.
1947 Communist government set up in Poland.
Romania abolishes monarchy.
1948 Communist coup in Czechoslovakia.
Communist people's republic established in Romania.
Communist forces set up North China people's republic.
North Korea becomes a communist republic.

1949   Communist régime established in Hungary.
          Communist republic of China established.

Even the election of a Labour government in Britain in 1945 was seen by some American officials as a communist takeover of the United Kingdom.

In addition, many countries became independent of their distant rulers (e.g. India and the newly-formed Pakistan became independent of Britain in 1947, as did Burma and Ceylon in 1948; Indonesia declared itself independent of Holland in 1949).

With all these new communist and/or independent countries forming, the political face of the world was in upheaval. To the US government it signified the beginning of a world dominated by communism, and to the Americans communism meant the USSR.

Thus began a propaganda war in which the US government told its people that the Russians were intent on ruling the world. It was therefore necessary (said the US) to prepare to defend the Free World (that which was not communist) against this threat.

In reality the 'threat' was a delusion. The Russians had enough problems internally. Frankly, they were not in a position, either economically or militarily, to invade America. They had just taken part in a war in which they had lost 20 million people. They were still struggling to put together a country that the Revolution of 1917 had dragged from the middle ages into the twentieth century almost overnight. There were severe differences between the USSR's leader, Stalin, and the other Russian leaders, which led to purges which caused many deaths, and economic hardships for the ordinary Russians. What the USSR wanted was to sort out its own problems. However, it saw the US government's build-up of weapons as a threat, and so it built up its own arsenals.

Things came to a head in 1960 when Cuba became a communist country under Fidel Castro. Cuba was not far from America, and the Americans saw this as the first real step towards an invasion of their country by communist forces. In 1961, the US government sent anti-Castro Cubans to take Cuba back from the communists, but the attempt failed.

In 1962 came the most knife-edged confrontation between the USSR and the USA. A Russian missile base had been established on Cuba. President John F. Kennedy of the USA demanded that it be dismantled. Premier Nikita Kruschev of Russia claimed that the missiles were only there to counter-balance the American nuclear missiles on the Russian borders in Western Europe. The Americans insisted that the missiles be removed, and sent warships to surround Cuba. For a few days, it looked as if the world was about to suffer its first full-scale nuclear war as other countries waited for the US government to attack the Cuban missile base, and the Russians to counter-attack.

Fortunately, the USSR withdrew its missiles, and the USA withdrew its warships. But what would the position have been if both countries had had different leaders? According to many historians, both Kruschev and Kennedy were 'sane liberals'. What if one of them, or both, had not been as sane, or as liberal?

This, in brief, is the 'Fear' part of the equation that has led to the build-up of nuclear weapons in the world. At the same time the 'Power' part of the answer was well advanced, with the influence of the USSR being felt not only in countries in Eastern Europe, but also in newly-emerging

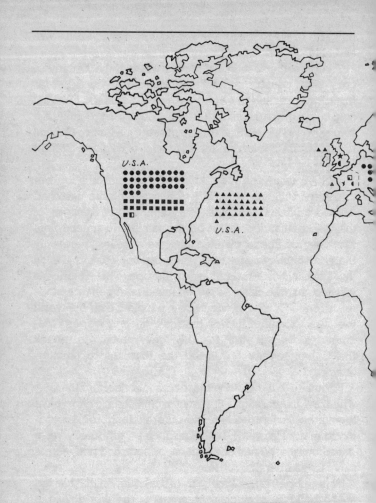

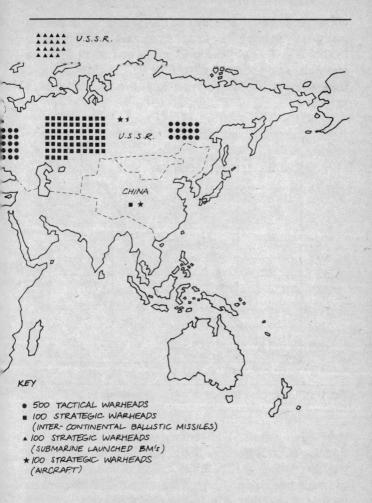

Fig. 20: The world's nuclear arsenals

countries in Africa. Equally, the USA was doing its best to make its influence felt by establishing military and financial footholds in Western Europe, Central and South America and Asia. And everywhere that the USA or USSR set up a base, strategic weapons were nearly always put in place to protect it.

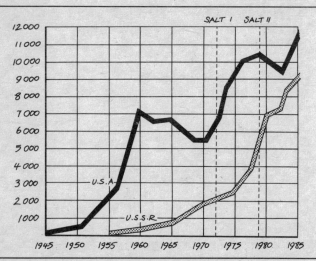

Fig. 19: The increase in nuclear weapons arsenals since 1945

In other words, both countries have been empire-building since 1945, spreading across the world in the same way that the Romans, the British, the Spanish and the French did centuries before. And it is this empire-building, based on fear of the other side, that has resulted in the northern half of our planet becoming a huge missile base, as you will see from Fig. 20.

This whole nuclear arms build-up is based on a policy known as Mutually Assured Destruction, or MAD for short. In my opinion, MAD is the correct way to describe it.

6 Nuclear Waste

What is nuclear waste?

Nuclear waste is the end result of nuclear fission. It is radioactive and is the rubbish left over after any nuclear work has been carried out, whether that work involves nuclear power, nuclear weapons or even hospital X-rays. Unlike ordinary rubbish, nuclear waste is dangerous for a very long time (see table on p 19).

In Britain, nuclear waste is classified under three main headings: high-level waste, intermediate-level waste and low-level waste. Generally speaking, the official nuclear industry classifications are that high-level waste includes spent fuel elements from reactors (e.g. uranium, which has a life of many thousands of years); intermediate-level waste includes equipment used in the reprocessing of nuclear waste, and therefore may include trace elements of uranium and plutonium; and low-level waste includes materials like gloves from medical research, old X-ray accessories etc. At least, that is the *official* line. The truth, as we shall see later in this chapter, is rather different.

The problem is that this nuclear waste has to be disposed of in some way. For many years it was dumped at sea in metal drums. However, it was discovered that once the drums were in the sea for long periods, they corroded and the radioactive waste leaked out and contaminated fish. Those who in turn ate the fish (including human beings) were then in danger of being contaminated by the radiation. There was also the danger of radiation being washed back on shore and contaminating beaches and coastal areas.

Because of these problems, and following action by Greenpeace, plus a refusal by the National Union of Seamen to carry nuclear waste out to sea and dump it, an International Conference of Maritime Nations held in London in 1985 decided to ban sea dumping of nuclear waste as unsafe. (The British Government's reaction to this ban was to protest about it, and then say that they would defy this ban whenever they could.)

To deal with the problem of how to dispose of Britain's nuclear waste, the Government set up an organisation called NIREX (Nuclear Industry Radioactive Waste Executive) in 1982. It consists of BNFL, CEGB, SSEB (South of Scotland Electricity Board), and UKAEA (United Kingdom Atomic Energy Authority).

As far as the high-level waste was concerned, NIREX originally decided to dump it in a deep mine owned by ICI at Billingham in north-east England, but such were the protests from the local community that they withdrew this plan. At the time of writing, they are waiting to see if their researches come up with another method of disposal. In the meantime, the high-level waste remains stored at the places that produce it (i.e. the nuclear power stations).

As far as the intermediate and low-level waste is concerned, NIREX decided to bury it in the ground in what they called 'shallow land burial dumps'. In order to persuade the public who lived in the areas they had selected as possible nuclear waste-dumps to accept this nuclear waste without too much protest, NIREX started up a publicity campaign. In the campaign they said that the waste would be mainly low-level waste from medical sources (X-rays, hospital gloves, etc). However, when they were asked to give evidence to the House of Commons Select Committee on the Environment Inquiry into Radioactive Waste, NIREX admitted that 'most of the radioactive waste arising in the United Kingdom comes from the nuclear generation of electricity. Smaller amounts come from the use of radioactive sources and radiation techniques in medicine and industry, and of course from the nuclear weapons programme.'

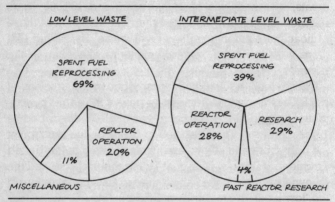

Fig. 21: The make-up of nuclear waste

In fact it was discovered that the ex-medical waste made up less than 16% of the nuclear waste NIREX wanted to bury in these shallow land dumps, the majority 84% would consist of:

- radioactive waste from power stations and nuclear reprocessing

- radioactive waste from Ministry of Defence nuclear weapons establishments (e.g. Aldermaston) radioactive waste from the reactors of Britain's nuclear submarines following reprocessing (particularly Polaris submarines, according to a Ministry of Defence memo).

This alleged 'intermediate-' and 'low-level waste', it was finally admitted, would include plutonium and uranium elements, with half-lives of 24,000 years and 162,000 years respectively. NIREX defended their classification of these radioactive elements as 'low-level waste' by saying that there would only be 'trace elements of plutonium in the waste'.

Those doubters who had already become suspicious of NIREX's reassuring statements in view of the above contradictions decided to investigate what NIREX described as 'trace elements'. They decided to check on the situation at the only existing shallow-land nuclear dump in the United Kingdom, at Drigg in Cumbria. These checks produced the following contradictions about the amounts of plutonium in the 'low-level' waste dump at Drigg:

1 A Senior executive of British Nuclear Fuels said on a Channel 4 television interview in December 1985 that there was *no* plutonium at Drigg.
2 The Minister for the Environment and NIREX said in 1986 that there was *some* plutonium at Drigg, but only 'trace elements'.
3 An official government report – the Flowers Report – said that by the end of 1976 the quantity of plutonium in low- and intermediate-level wastes at Drigg amounted to one-third of a tonne.

Once again the nuclear industry had been bending the truth about low- and intermediate-level wastes.

When the public realised the true nature of the waste to be buried at these shallow land dumps, protests started. To pacify the protestors, the Minister for the Environment, Kenneth Baker, announced in 1985 that he had changed his mind about the intermediate-level waste, and had told NIREX that only low-level wastes would be buried at the shallow dumps. So far so good, you might think. However a leaked memo from the CEGB showed that they planned to dilute the intermediate-level waste, and then reclassify it as low-level waste. In other words, it would still contain the same lethal long-lived radioactive elements like plutonium and uranium, but now it would have a different label on it telling the public that it was 'safe'.

> *'Mankind is not grown-up enough to be trusted with nuclear power and the safe handling of plutonium waste – the most lethal toxic man-made element on earth.'*
>
> Sir Kelvin Spencer, Chief Scientist to the Ministry of Power (1954–1959)

By 1985, NIREX had announced the four sites it wanted to investigate as potential nuclear waste dumps: they were Elstow in Bedfordshire, Fulbeck in Lincolnshire, South Killingholme in Humberside, and Bradwell in Essex. The sites had two things in common: they were all on clay, and they were all owned by the Government in one form or another, under the guise of the CEGB or the Ministry of Defence.

The official reason for the four sites being chosen was that clay was 'ideal for burying nuclear waste because the clay forms a natural barrier that stops it leaking out.' In fact the opposite is true. The bacteria in clay will corrode the mild steel drums that contain the waste, and the drums will leak within 40 years. When this was put to NIREX they admitted that it was true.

The reinforced concrete that NIREX propose to use to surround the dumps will crack within 50 years because the water in the clay will make the reinforcing rods expand. Clay is also unstable and cracks as it dries out in the summer months, and this will in turn lead to the concrete cracking. There is also no known concrete that will last as long as the nuclear waste the dump is intended to contain.

In other words, any shallow-land dump in clay will start to leak after only fifty years. As NIREX have admitted that the nuclear waste they plan to bury in these dumps will be lethal for at least 300 years (some of it will be lethal for many thousands of years), the whole idea of such dumps becomes ludicrous.

There is also the problem of transporting the nuclear waste from the places where it is produced (the nuclear power stations) to these dumps. NIREX calculate that there will be 30 to 40 lorry loads of dangerous radioactive waste going to these dumps each week, that is 2,000 lorry loads a year. The chances of an accident involving one of these lorries is enormous. A road accident (or a rail accident) involving nuclear waste would mean that the area around the accident would be contaminated by radiation, and we are then back to the Government's 'contingency plans' to deal with such an accident. As has already been shown, these plans would not deal effectively with a major accident at a nuclear power station, and they certainly would not be adequate to cope with nuclear road or rail accident.

Why, then, are NIREX considering these dumps as an option? And why have the British Government taken the unusual step of forcing the dumps on the local communities by means of a special order of parliament that says the dumps must go there whether the people who live there want them or not?

The answer is: money. It is cheaper to dig a hole in clay than it is to try to dispose of the nuclear waste in any other way. NIREX are also working on the 'out of sight out of mind' theory – once the waste is under the ground people will forget it is there.

Once again, this is an example of the nuclear industry's wishful thinking. In other countries, particularly the USA, where nuclear waste has been buried in underground dumps, the people who lived around those dumps are constantly reminded of it in the worst possible way. In the late 1940s nuclear waste was buried at Canonsburg in Pennsylvania. The US Government told the people at the time: 'This is only low-level radiation. Driving through it would give the same sort of radiation dose as smoking a cigarette.' In 1984 in Canonsburg the leukaemia rate in some streets was 4 cases in every 10 households. The national rate of leukaemia in the USA is only 3 to 4 cases in every 100,000 people.

In late 1957, an underground nuclear waste dump at Chelyabinsk in the USSR exploded. Winds carried a mixture of radioactive products and soil over an area of about one thousand square miles. In 1960, Russian scientists who visited this area found that it was so heavily contaminated that it had been declared 'forbidden territory'.

We are told that these problems were experienced early on in the life of the nuclear industry. We are assured by the nuclear industry that now, in the 1980s, things are different, that now the industry is aware of just how dangerous nuclear waste is. We are shown illustrations, like the one pictured here, of what a properly engineered shallow-land dump will look like.

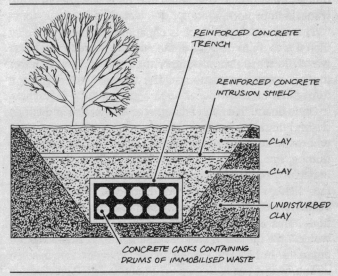

Fig. 22: Artist's impression of a nuclear waste dump

And just when we start to believe their claims that nuclear waste in the 1980s and the 1990s will be handled properly and carefully, we see a photograph taken in 1985 at the shallow waste dump at Drigg that shows how they handle nuclear waste 'properly' in the 1980s. It is just dumped off the back of a lorry into a big hole, and covered in soil – no concrete containment, nothing that remotely resembles their promises.

However, nuclear waste exists. If you do not want it buried in shallow-land dumps, then what can be done with it?

The answer from those of us in the anti-nuclear lobby is: we accept that the waste exists and that it must be dealt with in some way. What we do not accept is that shallow-land burial is the answer, and nor do most other 'nuclear' countries in the world.

We also think that one day, maybe in ten, maybe in twenty years, the scientists may come up with a safe answer to the problem of this waste. The nuclear waste research department at Harwell has an annual budget of £10 million to try to find an answer to the problem. At the moment they do not have a safe answer, nor do they have the technology to deal with the waste. Until they do, it should be stored above ground in dry storage, where it can be checked for leaks, at the places where it is produced, until such time as the scientists have found a way to deal with it. Storage of the waste should not present a problem because in 1985 Lord Marshall, the Chairman of the CEGB, was quoted as saying that the CEGB had space to store it for another hundred years.

The next move is to stop more of this lethal waste from being produced, and that means stopping nuclear power and nuclear waste reprocessing.

7 Uranium Mining

As we have seen, the key to the whole nuclear industry is uranium. Whether it's weapons or power, it needs uranium to get the whole thing going. Uranium is obtained by mining in much the same way that coal is mined, either by underground or by open-cast mining. The difference between coal and uranium is that uranium contains uranium 235, which, as we have already seen, is an unstable radioactive element that causes cancers and genetic defects. Another problem with uranium is that it gives off radon gas.

Radon is a colourless, odourless gas that occurs naturally. It is released in large amounts when uranium is mined and processed. It has been established by scientists and doctors that radon gas is a cause of lung cancer. It used to be thought that radon gas only affected those people directly involved in the mining of uranium (e.g. the actual miners), but recently it has been discovered that radon gas can affect everyone around the site of a uranium mine. The reason for this is because of the way that uranium is mined and processed.

As I said earlier, only certain parts of uranium are useful for nuclear fission, so this 'useful' uranium has to be extracted from the 'useless' uranium. This is done in the following way: after open-cast mining, the raw uranium ore is taken to a milling plant, where it is crushed into dust, and then mixed with water. The uranium is dissolved in acid, then separated from the undissolved solids and turned back into solid lumps of pure uranium oxide.

This milling process means that for every tonne of uranium ore mined, only about three kilos of uranium oxide are actually produced. This process means that there is a large amount of waste left over, and it is all contaminated with radioactive uranium. The waste is known as 'tailings', and it is estimated that these waste tailings still keep 80 per cent of the radioactivity of the uranium ore.

To give you an idea of how much radioactive waste is produced as tailings, the Ranger Mine in Australia, which produces 3,000 tonnes of uranium oxide a year, also produces one million tonnes of tailings in that same year.

The radioactive tailings are dumped into a huge pit known as a 'tailings dam'. There they produce radioactive pollution, which is spread about the surrounding area by wind and water – often these 'dams' leak into the water supply.

In New Mexico in the USA, the land on which the Navajo Indians live has the remains of 200 disused uranium mines spread over it, as well as millions of tonnes of radioactive waste. A study was made of the Navajo Indians who actually worked at a uranium mine in Cove, New Mexico. This study showed that out of 100 miners, 18 had already died of lung cancer caused by radon gas, and another 21 had developed cancerous malignancies. This gave a 'risk rate' of nearly 40 per cent for people working in that uranium mine.

To put the whole thing into perspective, a report by the United States Environmental Protection Agency said that up to 20,000 people die each year from lung cancer caused by radon gas. This includes miners, and those people living around the tailings dams who are affected by the radioactive contamination.

Why, then, are countries prepared to allow an industry such as uranium mining and milling, which kills so many people, to continue to exist?

The answer is money. Most of the world's uranium mines are in poor countries such as Namibia and Niger in Africa, and the workers are poor people who are glad of the money. The governments of these countries allow their people to work (and die) for the mining companies because they may be forced into it by stronger countries, or because they are paid by the mining companies, who are usually based in richer countries. For example, the world's biggest single producer of uranium is Rio Tinto Zinc, which is a British-based company.

In fact, the poor countries receive very little money for the uranium that is mined. Most of the money earned from uranium is kept by the mining company. To give you an idea of the large amounts of money involved, the Rossing uranium mine in Namibia, which is owned by Rio Tinto Zinc, made £25 million profit in 1985. This profit did not go to Namibia and the Namibians, but to the shareholders of Rio Tinto Zinc. What the Namibians got was the chance to work for very little money, and the chance to develop cancers from exposure to radon gas and radioactive uranium.

To sum up Part One of this book, it is my contention that the nuclear industry kills and damages people every step of the way, from uranium mining and processing, through nuclear power and weapons, right through to nuclear waste. What is worse, it kills and damages future generations. If we are going to save our planet and ourselves, then we must look for alternatives to nuclear energy.

Actually, we do not need to look for these alternatives, because they are already here, as I shall show you in the next chapter.

Part Two
The Alternatives

8 Energy from Renewable Sources

The myth that nuclear power is essential

According to the CEGB, without nuclear power we will 'freeze in the dark' – no heat, no light, no electricity for hospitals or for industry, etc. And because it is the 'experts' who are telling the public this, the public believes it. In fact this claim of freezing in the dark without nuclear power is nonsense.

For one thing, nuclear power provides only 20% of our electricity in the United Kingdom. This means that 80% of our electricity *does not* come from nuclear power. And if you look at the overall picture of how energy is used in this country (e.g. heating, transport, lighting, cooking, etc), electricity makes up less than 10% of this total.

In other words, the amount that nuclear power provides towards our total energy needs in the United Kingdom is 2% (20% of 10%), which means that 98% comes from other sources, a far cry from the 'freezing in the dark' claim of the nuclear industry.

It is also worth bearing in mind that the Department of Energy spends 90% of its research budget on nuclear power, while all the other sources (that together give us 98% of our energy) only get 10% for research.

Fossil fuels

So, what are these other energy sources?

At the moment, the major part of our energy is obtained

from coal, oil and gas. In 1983 they made up 94% of our energy (coal 36%, oil 34%, gas 24%). The problem with these fuels is that they are all fossil fuels, and will eventually run out. For example, in the case of oil and gas, we have consumed nearly 40% of the world's known supplies in the last fifty years. Coal, another fossil fuel, is also running out, although there are large reserves left in the world. Dr Michael Flood states that UK coal reserves should last for 'several hundred years at current rates of use'. However, coal presents us with another environmental problem: acid rain.

Acid rain is caused by the sulphur dioxide produced by coal-fired power stations, and it destroys forests and kills fish and plant life in lakes and rivers. Another problem with coal-fired power stations is that they release large amounts of carbon dioxide into the atmosphere, and this could increase the world's temperature. Although this may be thought of as a good thing in some parts of the world, in places like Iceland and Canada, where glaciers would melt, and in Africa, which already suffers from drought, the effects could be disastrous.

It is worth pointing out here that nuclear power is based on uranium, which is also a fossil fuel, and which will also run out – something else that will stop nuclear power from being 'the energy of the future'.

So, if fossil fuels are not the answer for the future, what is? The answer to that question is in two parts:

> 1. conservation of energy, and 2. the use of energy from renewable sources, e.g. the sun and water, which will not run out – at least, not in the foreseeable future.

Conservation

In the United Kingdom, 60% of our energy is lost because of the processes we use to produce that energy. For example, when converting coal to electricity, most of the heat and energy vanishes up the coal station chimneys and out into the atmosphere. Also, much of the heat in houses vanishes through cracks in doors, through uninsulated roofs, through windows, etc.

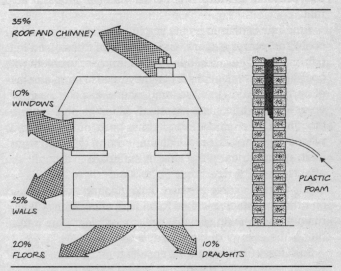

Fig. 23: How heat is lost from an uninsulated house

Proper conservation measures such as lagging a hot water tank, draught-proofing, insulating the loft, putting insulation into the outside walls of a house, and even putting in double glazing, can mean a saving of up to 40% in the amount of energy we need to produce.

Renewable Sources of Energy

Solar Power
When most people think of solar power they think of huge dishes trapping the sun's rays. However, the most immediate way to use the sun is 'passive solar heating'.

Passive solar heating It is called 'passive' because nothing spectacular happens, no amazing pieces of futuristic technology. All that happens is that special buildings are erected to catch the heat from the sun, and do not let it out again as quickly as most other buildings. In the new city of Milton Keynes, for example, there are estates of houses which have large double-glazed windows on the south-facing side of the house, and small windows on the north side. The large south-facing windows let in more heat from the sun than normal; the small windows on the north side are designed to keep the heat in the house.

At Wallasey in the north of England, there is a large school that has used passive solar heating successfully since 1961, without any need to use the conventional heating system that was installed at the same time, even in the middle of winter. The school has an enormous area of double glazed windows on its south side, and 50% of the school's heat comes into it as heat from the sun. The other 50% comes from the heat given off by the school's electric lights, and from the body temperatures of the pupils and teachers.

Solar Panels This is a diagram to show how a solar panel works:

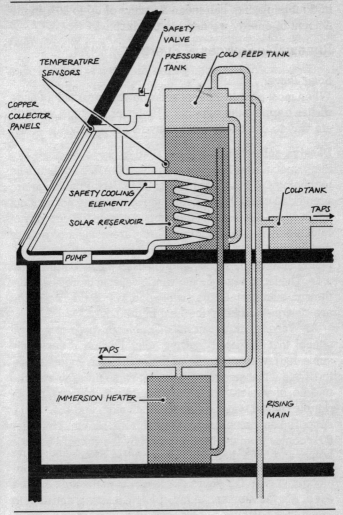

Fig. 24: How a solar panel works

Solar panels use the sun's rays to heat water, which is circulated around a house either as heat, or as a hot water supply. Solar panels can provide 40–50% of the hot water needed in a year.

On a larger scale, solar panels are already in use providing hot water for hospitals, hotels and swimming-pools.

Photovoltaic Cells These are electronic cells which use the sun's rays and convert them into electricity. At the

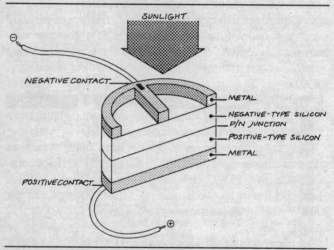

Fig. 25: A photovoltaic (solar) cell

moment the research is still in its early stages and in the United Kingdom there have been no major productions of electricity from this source, at least not in the amounts we need. For the moment, in my opinion, it seems sensible to use solar power for a) heat, and b) heating water, and to leave the production of electricity to other renewable sources that have already been successful, namely wind and water power.

Wind Power

The UK is one of the windiest places in the world, and yet we have not used the wind to produce electricity in very large quantities. This is a terrible waste of a natural and cheap energy source. The CEGB have said that up to 25% of our current electricity supply could be provided by wind power without major operational difficulties.

In Denmark, Sweden and the USA there are already wind-generators producing electricity at a cheaper rate than nuclear power. In America there are wind turbines generating as much as 2.5 MW (megawatts) of electricity. At Altamont Pass in California there is a whole bank of several thousand wind-generators producing electricity. At the time of writing, Denmark has 1,400 wind-generators producing electricity for their national grid.

The experiments so far in the UK have been in remote communities, such as South Ronaldsey in the North of Scotland, where a 22 KW (kilowatt) wind machine provides electricity for the local community. In the Orkneys there is a 300 KW wind-turbine providing electricity for about 150 houses. (See Fig. 26)

For those who think that wind-generators would be a blot on the landscape, I think it is worth pointing out that we have already become used to electric pylons all over the country. Besides, we do not need to place the generators on land, we can place them off-shore where the wind is that much stronger.

Water Power (Hydro Electricity)

Tidal Power The energy of the tides comes from the

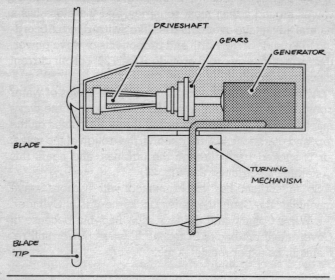

Fig. 26: How a wind turbine works

gravitational pull of the moon. Normally, tides are between one and three metres in height, but at river estuaries a funnelling effect is created. At the Severn Estuary, for example, this funnelling effect gives a range of up to 11 metres between high and low tides. A tidal barrage built at the Severn Estuary would provide about 5% of our electricity, the equivalent of *two* Sizewell B PWR nuclear power stations.

A tidal barrage is a dam that directs the incoming and outgoing tides past turbines which generate electricity. Tidal barrages could also be built at other river estuaries which have this same funnelling effect, for example, the River Mersey.

Wave Power Because it is an island, Britain has more waves around it than most countries, and therefore has a natural advantage over countries like Switzerland. According to Friends of the Earth and proponents of wave power, this form of energy could provide at least 25% of our current electricity supply.

A 'wave machine' works as follows: the waves of water compress flexible bags attached to the front of the wave machine. The air inside is pushed through turbines on the top, and electricity is generated. In the trough of the wave, air is sucked back through the turbines, and again this generates electricity. (See Fig. 27)

So far Britain has experimented with various wave machines: the Clam, the Salter Duck, the Bristol Cylinder, the Wave Column and the Lancaster Bag; but because of Government cut-backs in research, wave power still remains in its experimental stage.

Hydro-Electric Dams This is not a new idea; a hydro-power dam is actually recorded in the Domesday Book! Hydro-power was first used to generate electricity in the middle of the nineteenth century, and today it provides 2% of the United Kingdom's electricity supply. Large hydro-electric schemes in the north of Scotland already provide the cheapest source of power for the national grid.

Other Sources
Wind power, water power, and solar power are the usual sources of power that we think of when we talk of renewable (or alternative) sources of energy, but there are others not so well known, but just as important.

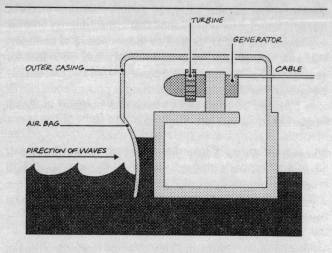
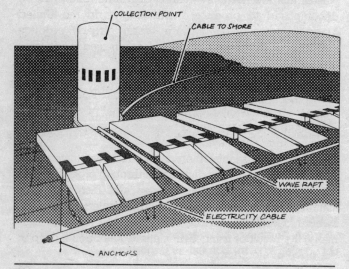

Fig. 27: Wave power generators – Top: The Sea Clam
Bottom: Wave Rafts

Biomass The phrase 'biomass' covers different kinds of energy. It includes living things that are an energy source, e.g. wood and plants, as well as the recycling of commercial and domestic wastes. In fact, biomass accounts for one-seventh of the energy used in the whole world, mainly because of the use of wood and dung in poorer countries. It also includes oil-seed rape, which can be used in diesel engines and turned into vegetable oil.

Recycling Waste Every year, over 30 million tonnes of household refuse is thrown away in the United Kingdom. If it was burnt to provide heat instead, it would provide the equivalent to 16 million tonnes of coal, or 5% of our annual energy consumption.

Methane Agricultural and animal waste (including human wastes that are normally flushed down the toilet) can be decomposed to produce methane gas. In the United Kingdom some of our sewage works already produce methane gas from sewage, and this is then used to generate electricity. (See Fig. 28)

CHP CHP stands for Combined Heat and Power. As we saw earlier in this chapter, in conventional power stations 60% of the energy is wasted because it disappears up the power station chimneys.

CHP stations use this otherwise wasted heat. Instead of it vanishing in smoke, it is distributed through pipes beneath the streets to surrounding houses and factories. CHP stations are already in use on a large scale in Denmark, Sweden, Germany and the USSR. In some areas of Germany, the result of using CHP stations has been a drop in the price of electricity.

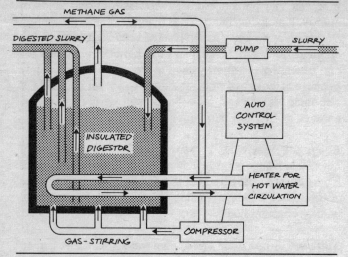

Fig. 28: A methane digester

Geothermal Deep below the earth lies geothermal, or 'hot rock' energy. It is tapped by drilling boreholes deep into the hot rocks and locating a basin called an 'aquifer'. Hot water comes up from the rocks and is used for heating. After the hot water has passed through the heating system, it is pumped back down to the aquifer. (See Fig. 29)

This system is already in use in France. In Paris, for example, a 1,700 apartment block has been using hot water from hot rocks for over 10 years.

Summing Up

There is some disagreement over the proportions that each of these different types of energy can contribute to our energy needs. Michael Flood of the Open University calcu-

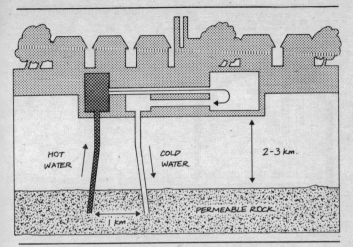

Fig. 29: How geo-thermal energy produces heat

lates that wind, biomass and geothermal energies would provide 85% of our renewable energy, the remainder coming from solar, tidal and hydro-electric. Other views are that wind, tidal and CHP would provide the bulk of our energy. Despite these differences, most researchers into renewable technology are convinced that between them these renewable sources can provide *all* our energy needs. The renewable sources have three advantages over conventional energy forms:

1   They are renewable and will not run out.
2   They are cleaner and less dangerous to our health and the health of all living things, and that includes our planet.

3   After the initial cost (e.g. building a wind-generator), they are cheaper to run.

This is why I, along with many other people, consider that renewable sources, not nuclear power, are the energy of the future.

9 Defence – the Non-Nuclear Alternative

Two main arguments are put forward for having nuclear weapons. They are:

1   Having a nuclear deterrent has stopped war from happening since 1945.

The figures do not support this. Since 1945 there have been 112 wars in the world, including the Korean War (1950–1953); the Vietnam War (1965–73); the Indian-Pakistan War (1965); the Soviet invasion of Afghanistan (1969 – the present); and the Falklands War (1982).

2   Having a nuclear deterrent stops other countries from attacking us.

In reality the opposite is true. Having nuclear weapons on our soil actually makes us a target for nuclear attack from an enemy because we pose a threat. Without nuclear weapons we would not pose that threat. Certainly no nation would want to conquer us using nuclear weapons, because after a nuclear attack our country would be uninhabitable as a result of the radiation. It is also worth remembering that the fact that Britain had a nuclear deterrent did not stop the Argentinians from invading the Falkland Islands.

Because nuclear weapons are very expensive (the budget for Britain's Trident nuclear missile programme is £22 billion), we have had to cut our conventional armed forces (the army, navy and air forces). This means that we are less well able to defend ourselves against attack from another country.

As a result, it is quite likely that we would start a nuclear war straight away if we were attacked by larger forces.

> *'The NATO doctrine is that we will fight with conventional weapons until we are losing, then we will fight with tactical weapons until we are losing, and then we will blow up the world.'*
>
> Morton Halperin, former US Deputy Assistant Secretary of Defence

Supporters of nuclear weapons also say that the weapons would never be used as first-strike weapons, but only as deterrents, to warn off a potential enemy. The problem with this is that it assumes that all the countries who have nuclear weapons are governed by sane leaders. History has shown that many leaders have been insane (for example, Adolf Hitler), and an insane leader certainly would not hesitate to use nuclear weapons as a first-strike weapon. For this reason alone there should be a ban on *any* country owning nuclear weapons.

It is not satisfactory to say that a particular country has a 'sane leader who would never use nuclear weapons unless forced to'. What happens after an election, or a coup, when that sane leader is replaced by a leader who is mentally unstable and eager to use whatever weapons come to hand?

There is also the risk of a nuclear war being started accidentally, as I pointed out in Chapter 5, yet another reason for banning all countries from owning these weapons.

> *'In an all-out nuclear war, more destructive power than in the whole of World War Two would be unleashed every second for the long afternoon it would take for all the missiles and bombs to fall. A World War Two every second, more people killed in the first few hours than all the wars of history put together. The survivors, if any, would live in despair amid the poisoned ruins of a civilisation that had committed suicide.'*
>
> *President Carter in his farewell speech to the American people, 15 January 1981*

For all these reasons, the only practical defence policy – a policy that is aimed at actually defending our country, as opposed to a policy that is aimed at attacking and conquering other countries – is one that eliminates nuclear weapons.

Apart from the fact that it enormously reduces the risk of our being a target for nuclear attack, this policy would also release billions of pounds to be put back into improving our real defences, such as high technology defence equipment, and a better equipped and better trained army, navy and air force.

It would also release money that can be put to use in making the fabric of our country stronger, by improving the health service, the education and welfare services, and by investing in industry. All of this would make Britain a healthier and more prosperous country, certainly much healthier and more prosperous than if we opted for a 'nuclear weapons and power' economy.

Glossary

acid rain rainfall contaminated with sulphur that has been emitted from coal-fired power stations. The sulphur can travel many hundreds of miles on the wind, and then mixes with falling rain. Acid rain kills trees, and fish and vegetation in lakes.

AGR Advanced Gas Cooled Reactor. A nuclear reactor where the coolant is carbon dioxide gas rather than water. (See 'coolant'.)

alpha particles heavy nuclear particles given off in radioactive decay. Not very penetrating.

aquifer A hollowed-out underground basin that contains water.

atom the smallest particle taking part in a chemical action.

atomic number the number of protons in the nucleus of the atom.

BWR Boiling Water Reactor. A reactor where the coolant is water, which is turned to steam.

bequerel one atomic disintegration per second. It is a measure of the amount of radioactivity, so if milk has 2000 bequerels per litre, this means that in each litre of milk 2000 atoms break up per second.

beta particles electrons emitted from the nucleus of an atom in certain types of radioactive decay.

biomass material made from living things that stores solar energy (e.g. when wood is burnt as fuel the sun's energy stored in it is released).

BNFL British Nuclear Fuels Ltd.

CEGB Central Electricity Generating Board.

chain reaction the process in which one nuclear occurrence causes the same thing to happen in another atom.

CHP Combined Heat and Power. A system whereby the steam from a power station heats surrounding buildings (houses, factories, etc.), instead of being lost into the atmosphere up the power station chimneys.

control rod a rod of steel or aluminium containing material which absorbs neutrons. These are used to maintain a nuclear reactor at a certain level.

coolant gas or liquid that is pumped through the core of a nuclear reactor to cool it. This coolant then becomes hot and produces steam.

core the heart of a nuclear reactor, containing the fuel elements, the moderating material, and the control rods.

curie a unit of radioactivity, named after Marie and Pierre Curie. It is equal to 37,000,000,000 bequerels. This term is really no longer used, but it is sometimes still referred to.

EMP Electro-Magnetic Pulse.

enriched uranium the very fissile uranium 235 that remains after most of the less fissile uranium 238 has been extracted from uranium oxide.

FBR Fast Breeder Reactor. A nuclear reactor that produces more fissile material than it consumes.

fallout radioactive dust and other matter that falls back to earth from the atmosphere after a nuclear explosion.

fissile material material that is capable of undergoing fission when it is struck by a slow neutron.

fission the splitting of a nucleus. This process can be spontaneous, or it can be caused by the impact of a neutron striking the nucleus.

gamma rays very penetrating rays given off by radioactive material. Thick barriers of lead or concrete are needed to stop them.

gray a measurement of the energy given to tissue by radiation.

half-life the length of time it takes for the level of radioactivity of an isotope to be halved.

Hibakusha the Japanese term for those who still suffer from the effects of the atomic bombs dropped on Hiroshima and Nagasaki.

ICBM Inter-continental Ballistic Missile. A missile capable of travelling from one continent to another.

kiloton the equivalent of one thousand tons (1016 tonnes) of TNT.

MAD Mutually Assured Destruction. A policy whereby if one major power starts a nuclear war, the combined nuclear arsenals of the combatants will ensure that they will destroy each other.

This is supposed to be a deterrent.

magnox magnesium alloy used to case the uranium in reactor fuel elements.

Magnox reactor a nuclear reactor in which the heat is extracted by carbon dioxide gas.

megaton the equivalent of one million tons (1,016,000 tonnes) of TNT.

meltdown this would occur if the core of a nuclear reactor overheated to such a degree that the fuel melted into one large lump. This lump would be beyond control and would melt into the earth.

millisievert 1/1000 of a sievert.

moderating material this material moderates the speed at which neutrons travel. (E.g. carbon is a moderating material.)

NATO North Atlantic Treaty Organisation. A military alliance between the following countries: Belgium, Canada, Denmark, France, Greece, Iceland, Italy, Luxembourg, Netherlands, Norway, Portugal, Spain, Turkey, UK, USA and West Germany.

neutron one of the two particles of the nucleus. It is heavier than a proton. It has no electrical charge.

NIREX Nuclear Industry Radioactive Waste Executive.

NORAD North American Aerospace Defense Command.

NRPB National Radiological Protection Board.

nucleus the centre of an atom, made up of protons and neutrons.

photovoltaic cell usually called a 'solar' cell. When photons (sunlight) strike the silicon chips it generates electricity.

proton one of the two particles of a nucleus.

PWR Pressurised Water Reactor.

rad this stands for Radiation Absorbed Dose. It is equal to 1/100 of a gray.

radiation the particles and rays given off by an unstable atom.

radon gas a radioactive gas given off by uranium.

reactor the place where the nuclear reaction takes place (i.e. where the radioactive elements in the fuel are split by neutrons).

rem this stands for Roentgen Equivalent Man. For beta and gamma radiation and X-rays, one rem is equal to one rad. One rem equals 1/100 of a seivert.

roentgen a measurement of energy given to non-biological matter.
sievert a measurement of damage done to tissue by radiation. One gray of beta or gamma radiation absorbed gives a dose of one sievert. Alpha radiation is more damaging, and one gray of alpha radiation gives about 20 sieverts.
spent fuel the fuel elements after they have been subjected to nuclear fission. They are still radioactive.
tailings the waste material left over after uranium oxide has been extracted from raw uranium.
x-rays produced when electrons bombard a metal target.

List of Useful Organisations

Friends of the Earth (UK) 377 City Road, London EC1V 1NA. An international organisation devoted to cleaning up our planet. Concerned with all forms of pollution, acid rain and lead, as well as nuclear. As well as gathering and publicising information, it lobbies politicians for change.
Greenpeace (London), 36 Graham Street, London N1. An international organisation, like FoE, devoted to cleaning up our planet, but more activist in its approach (e.g. sailing ships into nuclear test zones to stop weapons being tested).
Campaign for Nuclear Disarmament, 22 Underwood Street, London N1. A British organisation devoted to the banning of all nuclear weapons, and particularly those on British soil. It has a youth group, Young CND.
European Nuclear Disarmament, 11 Goodwin Street, London N4 3HQ. A European organisation, linked with CND, devoted to banning nuclear weapons in both Western and Eastern Europe.
Centre for Alternative Technology, Llwyngwern Quarry, Machynlleth, Powys, Wales. Established in 1974, it is a community that not only researches but relies on alternative technologies. They generate their own electricity, produce as much of their own food as possible, and recycle their waste. The Centre is open for visitors daily from 10am to 5pm. (Telephone: Machnylleth 2400)
NATTA (Network for Alternative Technology and Technology Assessment), c/o Alternative Technology Group, The Open University, Walton Hall, Milton Keynes, Bucks. An organisation of academics who are researching and assessing renewable sources of energy on a worldwide scale.
The Green Party, 36 Clapham Road, London SW9 0JQ. The British wing of the international political Green movement. The only totally anti-nuclear party in British politics.
SERA (Socialist and Environment Resources Association), 9 Poland Street, London WC1V 3DG. The anti-nuclear group within the Labour Party, working to get the Labour Party to adopt a non-nuclear power policy.

Research and Further Reading

'The Medical Effects of Nuclear War' (The Report of the British Medical Association's Board of Science and Education) (John Wiley & Sons 1983)

Phil Bolsover, *Civil Defence: The Cruellest Confidence Trick* (CND 1980)

Stewart Boyle, *Reach for the Sun* (Friends of the Earth)

Stewart Boyle (ed), *Critiques of the Black Report* (Friends of the Earth)

Rosalie Bertell, *No Immediate Danger* (The Women's Press, 1985)

Peter Bunyard, *Nuclear Britain* (New English Library, 1981)

Rénée Chudleigh & William Cannell, *The Gravedigger's Dilemma* (Friends of the Earth)

Mike Cross, *Wind Power* (Franklin Watts/Aladdin Books, 1985)

Michael Flood, *The Potential For Renewable Energy* (Alternative Technology Group, The Open University, 1986)

Owen Greene/Barry Rubin/Neil Turok/Philip Webber/Graeme Wilkinson, *London After the Bomb* (Oxford University Press, 1982)

Nigel Hawkes/Geoffrey Lean/David Leigh/Robin McKie/Peter Pringle/Andrew Wilson, *The Worst Accident in the World* (Pan/Heinemann, 1986)

Robin McKie, *Nuclear Energy* (Macdonald, 1984)

Paul McClory: *Alternative Energy* (Wayland 1985)

C. A. Mann, *Nuclear Fuel* (Wayland, 1979)

Sherry Payne, *Wind And Water Energy* (Blackwell Raintree, 1983)

Danny Powell & Lindy Williams, *Will They Thank Us for This?* (The Green Party, 1986)

Alan Roberts & Zhores Medvedev, *Hazards of Nuclear Power* (Spokesman, 1977)

Andrew Wilson, *The Disarmer's Handbook* (Penguin, 1983)

Ground Zero, *Nuclear War* (Methuen, 1982)

Acknowledgements

I would also like to acknowledge the enormous help that the following people gave me. Without their help and assistance this book would never have been possible:

Stewart Boyle, Bernadette Vallely, and all at Friends of the Earth (to whom this book is dedicated); George Pritchard and all at Greenpeace; Steve Wilson; Gisela Probst, Collum Grant; Caryl Myrull; Duncan Eldridge; Alison Randall; and Rosemary Canter. My special thanks to them.